Estimate
with Backpack Bear

by Pam Ferguson

Illustrated by Faith Gowan
and the Starfall team

Starfall Education Foundation
P.O. Box 359, Boulder, CO 80306

Printed on recycled paper.   Library of Congress Control Number: 2015959553   ISBN: 978-1-59577-217-6

How to use this book

In mathematics we almost always emphasize arriving at a precise answer as the goal of a given process. This may explain why the important skill of estimation can be a challenging concept for some children. Young mathematicians want to be right. When they see 17 objects, answering "about 20" is not good enough.

The ability to estimate is a valuable and necessary skill. It demonstrates good number sense and enables children to evaluate whether their solutions to math questions are reasonable. Estimation focuses our attention on the available information and engages our logic to come to a reasonable response.

The estimation activities in this book guide children to understand that when estimating, you are right when you make a smart guess based on the available information rather than calculate a precise answer. In many instances the estimation exercise has more than one "correct" response. This will foster dialogue about the reasonableness of each answer.

Come and play a game with me
We'll learn a handy tool.
First think and guess, then estimate
It's a very simple rule!

When we don't know just how far it is
Or how much time a task will take
We'll use the clues around us,
Make a smart guess, and estimate!

In everyday life, from time to time
We need an answer in a hurry.
We estimate to answer fast
Close enough is right—don't worry!

Look at these dots.
Don't count them!
Are there more **purple** dots or **green** dots?

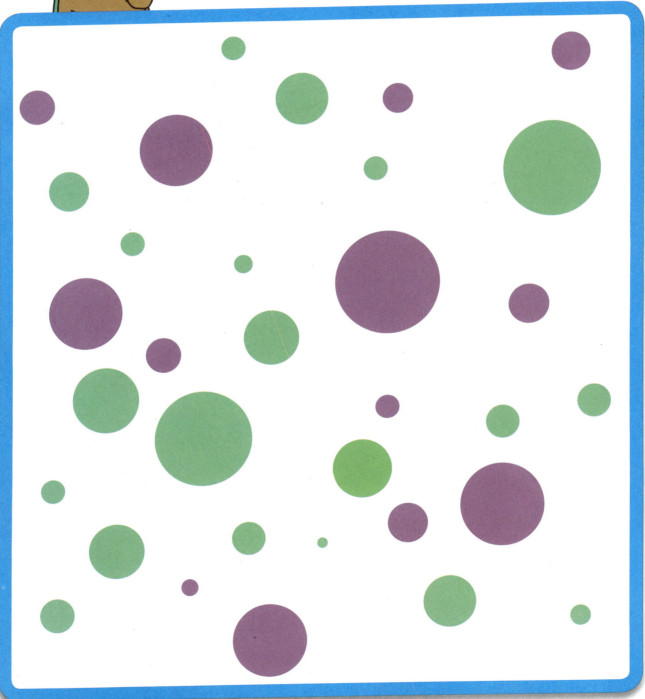

This is the number **8.**
Is it closer to **3, 12,** or **9**?

Look at these shapes.
Are there more circles or more squares?
No counting!

What a beautiful tree.
Do you think there are more leaves
on the ground or on the tree?

Let's measure how high the door is in our room!
Will it take more craft sticks or paper clips to
do the job?

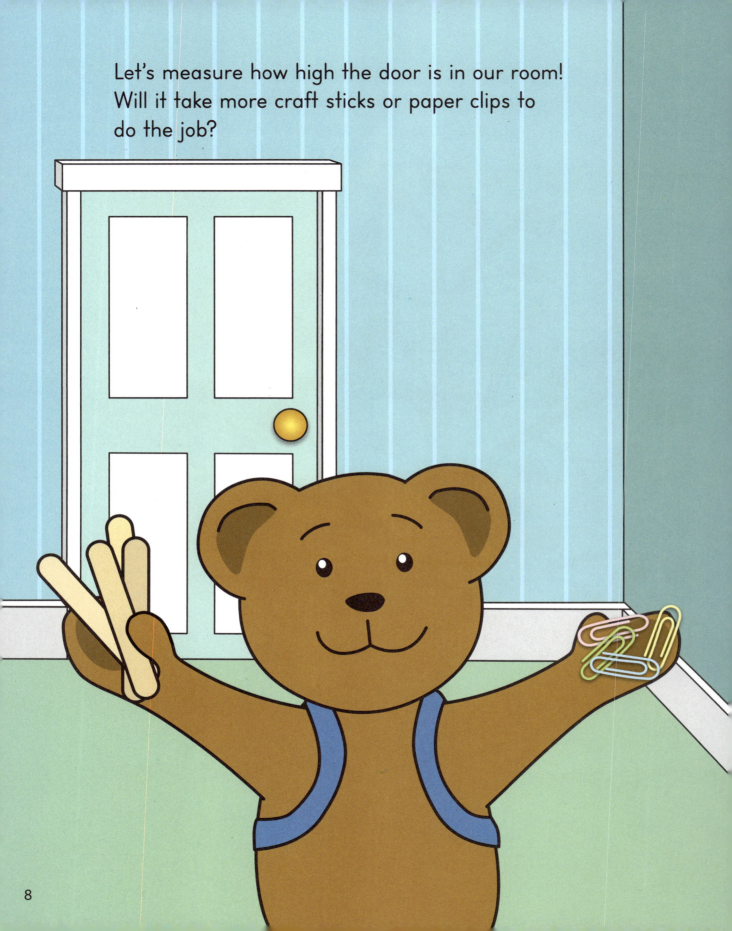

Do you think more caterpillars or more ladybugs will fit on this leaf?

This is an awesome picture of the night sky.
There are too many stars to count.
Do you think there are

About **30** stars?

Almost **50** stars?

Or more than **100** stars?

Look at all of these shapes.
Don't count them!
Are there more **2D** or more **3D** shapes?

This basket is filled with eggs.
How many do you think there are inside?
 Less than **10** eggs?
 About **50** eggs?
 Closer to **20** eggs?

This is the number **30**.
Is it closer to **40**, **10**, or **35**?

How many little hearts would fit
inside the large heart?
Do you think it would be
      Less than **10**?
      About **20**?
      More than **50**?

Which would take more time?
Going to the toy store on a bike or in a car?

How many grapes do you think
your friend can eat in a minute?
Less than **10**?
About **15**?
More than **20**?

This garden is full of vegetables!
There may be more of one vegetable than the others.
Which vegetable has produced the most?

Look at this harvest! Did Backpack Bear pick
more of one vegetable than the others?

How many fish will Tim catch in an hour?
Less than **5**?
About **8**?
More than **10**?

If we all worked together, how long do you think
it would take to clean up after playtime?
Between **1** and **5** minutes?
About **10** minutes?
More than **30** minutes?

How many more frogs will fit in this pond?
Less than **3**?
About **6**?
More than **12**?

Imagine today is your birthday!
You invited 20 of your best friends to your party,
but even more friends might come.
How can you make sure there are enough
cupcakes for everyone?

Make less than **20**?

Make **20**?

Make more than **20**?

How many pages do you think are in this book?

Less than **20**?

About **30**?

Closer to **50**?

Do you like to exercise?
How many jumping jacks do you think
you can do in one minute?
　　　Less than **10**?
　　　About **20**?
　　　More than **30**?

Do you think this car weighs
less than you?
About the same as you?
More than you?

How long is one minute?
Close your eyes.
When you think one minute is up, open them.

How many coins will fit inside this piggy bank?
About **10**?
About **15**?
More than **20**?

Let's fill this bathtub with beach balls!
How many do you think will fit inside?
Between **0** and **5**?
6 to **10**?
More than **10**?

There are 12 months in the year.
Let's name them.
What month is it now?
How many more months is it until Thanksgiving?

1 to **4**?

5 to **8**?

More than **9**?

Let's go swimming!
Guess the temperature of the water.
Do you think it is
      More than **100** degrees?
      Around **80** degrees?
      Under **50** degrees?

How many minutes do you think
it would take to read this book?
Less than **5** minutes?
About **10** minutes?
More than **20** minutes?

This puzzle is missing some of its pieces.
How many do you think are missing?
About **2**?
Less than **1**?
More than **4**?

You want to buy three toys.
The first toy costs 10 cents.
The second toy costs 40 cents.
The third toy costs $1.10.
How much money should you bring to
the store to buy all of these toys?

      I cent to **$1.00**?

      $1.10 to **$2.00**?

      Over **$20**?

Let's open a lemonade stand!
We have one gallon of lemonade.
These are our cups.
How many cups of lemonade do you think we can sell?

More than **100**?

About **6**?

More than **10**?

How many steps would it take to walk
from this room to the bathroom?
About **20**?
Around **40**?
More than **50**?

Your shoelace has broken.
You will have to replace it.
Do you think the shoelace should be
10 paper clips long?
30 paper clips long?
Close to **100** paper clips long?

Yum! Pizza!
How many pizzas do we have?
About **2**?
Closer to **3**?
More than **3**?

Someone has eaten some of this birthday cake.
How much has been eaten?
     More than half of the cake?
     About half of the cake?
     Less than half of the cake?

Look at these numbers:

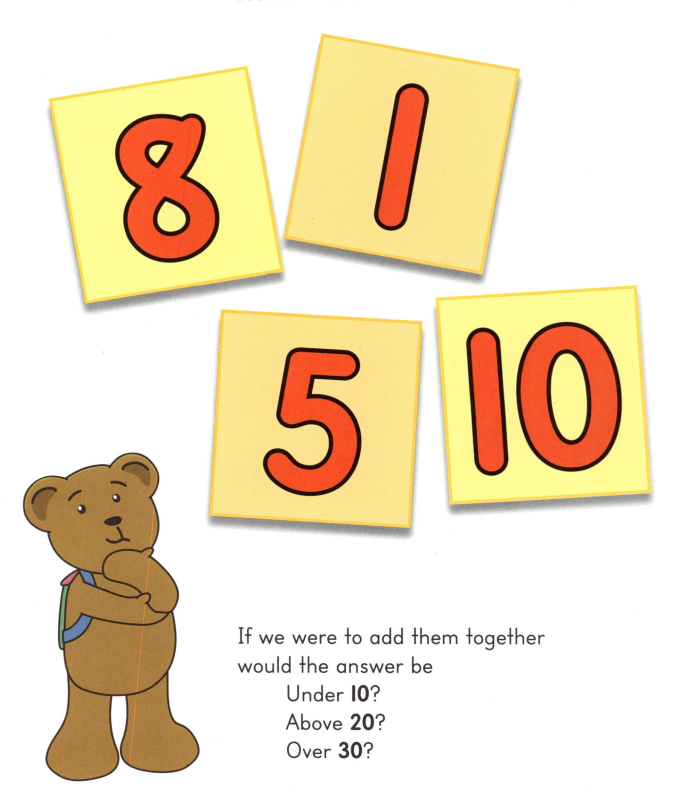

If we were to add them together
would the answer be
     Under **10**?
     Above **20**?
     Over **30**?

How many sticks do you see?
Closer to **16**?
Closer to **30**?
Closer to **7**?

Index by Skill